Primary Geography

Pupil Book 3 Investigation

Stephen Scoffham | Colin Bridge

Planet Earth

Landscapes
The Earth's surface — 2-3
The shape of the land — 4-5
Investigating landscapes — 6-7

Water

Water around us
A wet planet — 8-9
The effects of water — 10-11
Recording water — 12-13

Weather

Weather worldwide
Different types of weather — 14-15
Living in hot and cold places — 16-17
Sunshine matters — 18-19

Settlements

Villages
A village community — 20-21
Different types of village — 22-23
Investigating villages — 24-25

Work and Travel

Travel
Ways of travelling — 26-27
Finding your way — 28-29
Routes and journeys — 30-31

Environment

Caring for the countryside
Wildlife around us — 32-33
Protecting wildlife — 34-35
Improving our surroundings — 36-37

Places

Scotland — 38-43
France — 44-49
South America — 50-55
Asia — 56-61

Glossary — 62

Index — 63

Unit 1 Landscapes

Lesson 1: The Earth's surface

> What is the Earth's surface like?

Key words

continent
Earth
landscape
planet
space

The Earth is one of eight planets that spin around the sun. From far out in space the Earth looks quite small. You can see it is round like a ball.

The Earth is a very special planet. There is just the right mix of rock, air and water for animals and plants to live. No other planet is quite the same.

Water covers most of the Earth's surface. There are many different seas and oceans. The Pacific Ocean is the largest. The land is divided into great blocks called continents. Some of the land is very flat. In other places there are mountains, forests and deserts.

The surface of the moon.

Discussion

- Which of the photographs on this page shows the Earth?
- What three clues helped you decide?
- Talk about how people might use the landscape shown in the photograph of the Earth.

The surface of the Earth.

Unit 1 • Landscapes

A picture of the Earth from space made from satellite images.

Data Bank
- The Earth is over 4000 million years old.
- Beneath the surface, the Earth is made of red hot rocks.
- The Earth takes a year to spin around the sun.

Mapwork
Find a globe and turn it to the view in the space photograph. Name some of the places you are looking at.

Investigation
Make a list of the colours you can see in the picture of the Earth on this page. Say what you think each one shows.

Unit 1 • Landscapes

Lesson 2: The shape of the land

Are all landscapes the same?

The shape of the land is called the landscape. Landscapes form very slowly over millions of years. Mountains are worn away by snow, ice, wind and rain. In other places, the land is rising. Geographers study how these changes happen.

Mountains are steep and rugged places. There is only a little soil for plants and the weather is often bad.

▼ The Himalayas in Asia.

Discussion

- Which landscape would be best for (a) rock climbing (b) walking?
- Why don't trees grow in every landscape?
- Which landscape is most similar to the place where you live?

Mapwork

Make a map or a model of an imaginary island with a number of different landscapes.

▲ A plateau in Chile, South America.

A plateau is a flat area found high up in the mountains. The weather here is often windy.

Unit 1 • Landscapes

Hills are not as high as mountains but can have steep slopes made by rivers. There is enough soil for grass and trees to grow.

▼ Hills and valleys in France.

Islands are areas of land which are surrounded by water. They are often found in groups.

▼ A Canadian island in the Pacific Ocean.

Key words

coast
hill
island
landscape
lowland
mountain
plain
plateau
valley

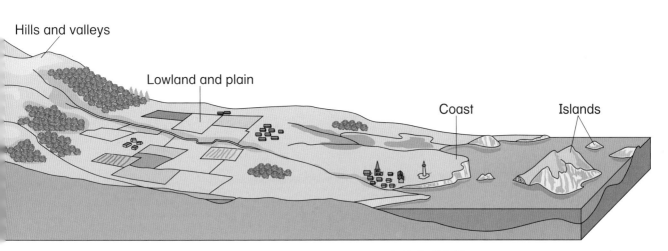

▲ Fields in Oxfordshire, southern England.

Lowlands and plains are flat landscapes. Many people live in lowland areas because they have the best farmland.

▲ The rocky coast of South Wales.

The coast is where the land meets the sea. Some coasts are rocky; others have sandy beaches, marshes or swamps.

Investigation

Start to make a geography notebook. Write down four landscape words and draw pictures to go with them.

Unit 1 • Landscapes

Lesson 3: Investigating landscapes

Key words

British Isles
Grampian Mountains
River Thames
Snowdonia

What is the landscape like in the British Isles?

The British Isles are made up of mountains, hills and lowlands. Most of the mountains are in the north and west. There are lowlands in the south and east. The rocks which make up the landscape date back as much as 700 million years.

Look at the map carefully. Can you find where you live? Do you live in the mountains, hills or lowlands?

Key
- Over 500 metres
- 200-500 metres
- 0-200 metres

Data Bank
- The Grampian Mountains have three or four months of snow each year.
- Some lowland areas are below sea level and protected by walls.

Scale
0   100   200   300 km

Discussion
- Are the British Isles mostly hilly or flat?
- Which colour on the map shows where most people live?
- Which sea is closest to where you live?

Mapwork
Working from the map make a list of (a) mountain ranges (b) rivers.

Unit 1 • Landscapes

A local enquiry

At St Mary's School the children did a project about their local landscape using a map and an aerial photograph of their area. They imagined what the landscape would look like without any buildings. First they listed all the landscape features such as hills, valleys and streams. Then they made their own map of the area showing all the features on their list. Finally the children wrote a short report about what their map showed. You could do a similar project about your local landscape.

The children made a large plan.

They linked a list of landscape words with a list of descriptions.

Investigation
Download some pictures of different landscapes or cut them out from magazines. Write some sentences about each landscape for a class wall display.

Summary
In this unit you have learnt about:
- the surface of the Earth
- different landscape features
- how to study the landscape.

▲ Aerial photograph of the local landscape.

Unit 2 Water around us

Lesson 1: A wet planet

Where do we find water?

Almost three-quarters of the Earth's surface is covered by water. Most of it is in the seas and oceans.

There is also a lot of water on the land. There is water in ponds, streams, rivers and lakes.

Wherever you live, water is all around you. It is in the rocks and soil under your feet. Water is also in the air and forms clouds above your head.

Key words

cloud
glacier
iceberg
lake
pond
river
stream

▼ The Victoria Falls are on the River Zambezi in Africa.

Discussion

- What are the different places where water is found?
- Where could you see, feel and hear water in your area?
- Which is the most useful form of water in people's lives?

Unit 2 • Water around us

Clouds are made up of millions of tiny water droplets.

Liquid

When water droplets join together it starts to rain.

Gas

An invisible gas called water vapour fills the air around us.

Solid

If the air is very cold, the water droplets turn into snow or ice.

Rain fills the dips and hollows in the land to make lakes.

Snow can form into rivers of ice called glaciers.

Rivers run down slopes to the sea.

Icebergs are huge blocks of ice floating in the sea.

All water ends up in the sea.

Investigation

Make a drawing or plan of a house. Show in which rooms you would find (a) ice (b) running water (c) water vapour (steam).

Mapwork

Using an atlas make a list of lakes around the world.

Unit 2 • Water around us

Lesson 2: The effects of water

Key words

flow diagram
pond
soil
water

Why is water important?

Without water everything would die. Trees and plants need water from the soil. People and animals use water for drinking and keeping themselves clean. Fish and many other animals need to live in water.

Birds visit ponds to drink and to clean their feathers.

Apples and other fruit are mostly water.

Most plants need lots of water to grow well.

Fish live in water. If ponds dry out they die.

Tree roots take water from deep down in the soil.

Worms and insects need water to digest their food.

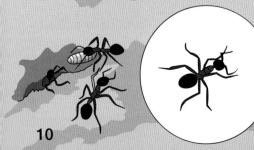

Ants can live longer without water than any other animal.

Data Bank

- Water makes up about three-quarters of our body weight.
- Each person in the UK uses about 150 litres of water a day.

Unit 2 • Water around us

Using water

Drinking and cooking.

Washing and cleaning.

Watering plants.

Travelling.

Discussion
- Why do living things need water?
- What are the main ways people use water?
- Is water in the UK used in similar ways to people in the photographs?

Mapwork
Make a flow diagram with the word 'water' in the middle. Give examples of drinking, cooking, washing and so forth round the edge.

Investigation
Look at the picture of the pond. List plants and animals which (a) use water from the pond (b) use underground water.

Unit 2 • Water around us

Lesson 3: Recording water

How is water shown on maps?

Water is shown in blue on maps and plans. Streams, rivers, marshes, lakes, seas and reservoirs are some of the things which are marked.

Key words

co-ordinates
lake
marsh
reservoir
river
stream

Mapwork

Work out the letter and number co-ordinates for:

River Blackwater

Mersea Island

Abberton Reservoir

Geedon Saltings.

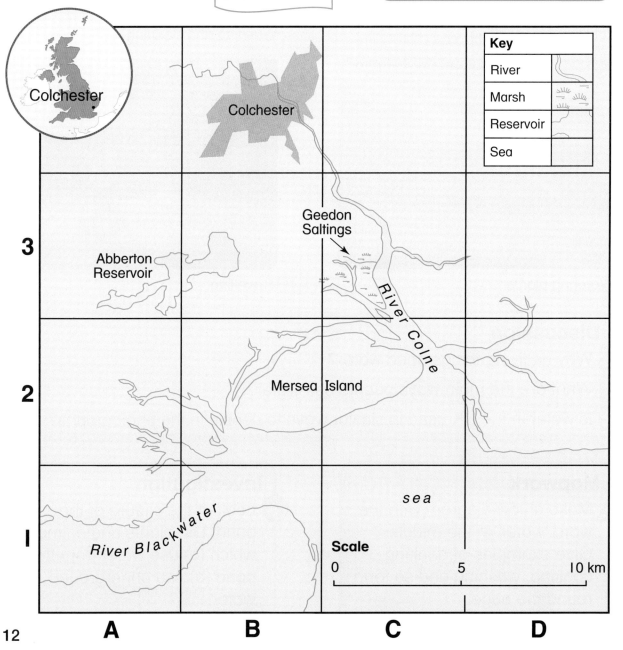

Unit 2 • Water around us

Closer to home

When rain reaches the ground it flows down hills and slopes. It can:

- run into streams and rivers
- stand in ponds and puddles
- soak into the soil
- go back into the air.

The rain falls from the clouds.

It runs off the roof and into the gutter.

The rain is used by trees and plants.

The rain collects in ponds and puddles.

It flows down the downpipe.

It goes down the drain.

The rain soaks into the ground.

The rain runs into streams.

Investigation

Make a survey of where rainwater goes in your school grounds. Write a report or draw a picture like the one on this page.

Discussion

Which are the wettest and driest places around your school?

How does water disappear from puddles, ponds and lakes?

Can you find any place names that involve water from a map of your area?

Summary

In this unit you have learnt:

- about the different forms of water, where water is found, and where it goes
- why water is important to animals, plants and people
- how water is shown on maps.

Unit 3 Weather worldwide

Lesson 1: Different types of weather

Is the weather the same all over the world?

Key words

climate
desert
North Pole
polar lands
rainforest

There are different types of weather around the world. Some places are very hot. Others are cold. In some places it rains a lot. In others it is very dry. The pattern of weather over a number of years is called the climate.

Data Bank
- The Sahara Desert is over 5000 km across.
- The North Pole and Antarctica are covered by vast sheets of ice.
- The Amazon rainforest is nearly the size of Europe.

Rainforest
In the rainforest the climate is hot and wet. Huge numbers of plants and animals live in the trees. Wide rivers flow through the forest.

Discussion
- Compare the climate in the desert, rainforest and polar lands.
- What would you put in a survival kit for each place?
- What animals might live in each region? Which has the hardest job to survive and which the easiest?

Unit 3 • Weather worldwide

Desert
In the desert the climate is very dry. Plants and animals have to live on very little water. Some deserts are very hot.

Polar lands
Polar lands have the coldest climate on Earth. In some places snow and ice last all year. Very few plants and animals can live there.

Mapwork
Add drawings of the rainforest, desert and polar lands to your geography notebook. Write a sentence about each one.

Investigation
Using the internet find out about extreme weather records from around the world.

Unit 3 • Weather worldwide

Lesson 2: Living in hot and cold places

How do people live in hot and cold places?

Key words
date palm
desert
market
oasis
polar lands
temperature

In the past, people found it hard to survive. They had to grow all their own food, make their own clothes and keep their own animals. They used whatever materials they could find to build houses. If the weather was too hot or too cold their crops or animals could die.

Today, people can live almost anywhere in the world. Electricity, machines and modern homes help us to cope with difficult weather. Food can be delivered by lorry or by plane. However the weather still affects us.

Discussion
- What are the main problems of living in the desert and polar lands?
- How does the weather affect your life?
- How does very hot or very cold weather affect people?

Living in polar lands

In Greenland some people earn a living from fishing. What do you think is happening in these two photographs?

Unit 3 • Weather worldwide

Living in the desert

There are only a few places in the desert where people can find water. These places are called oases. Some oases are small villages. Others are towns and have important markets.

▼ An oasis in Morocco.

Many desert houses have flat roofs because there is very little rain.

Camels used to carry heavy loads but they have been replaced by cars and trucks.

Food crops, like wheat, grow well in the desert when they have enough water.

All life depends on water in the desert.

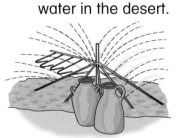

Date palms grow in the heat of the hot desert sun.

Mapwork
Make a picture map or a model of an oasis.

Investigation
Draw two things in a hot desert and two things from polar lands that are different to the UK.

Unit 3 • Weather worldwide

Lesson 3: Sunshine matters

Why are some places hot and other places cold?

At the Equator the sun rises high in the sky. The sun's rays fall straight on to the Earth and heat it up.

Near the North and South poles the sun is always low in the sky. There are long shadows and the air never gets warm.

Key words

Equator
North Pole
South Pole
reservoir

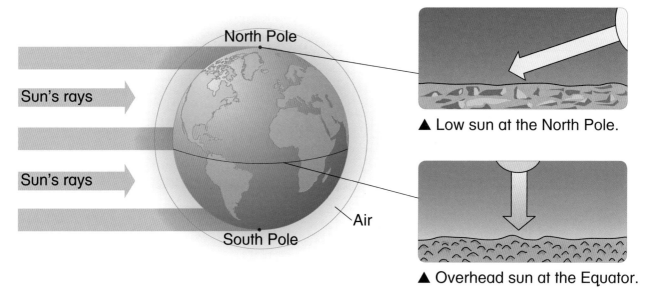

▲ Low sun at the North Pole.

▲ Overhead sun at the Equator.

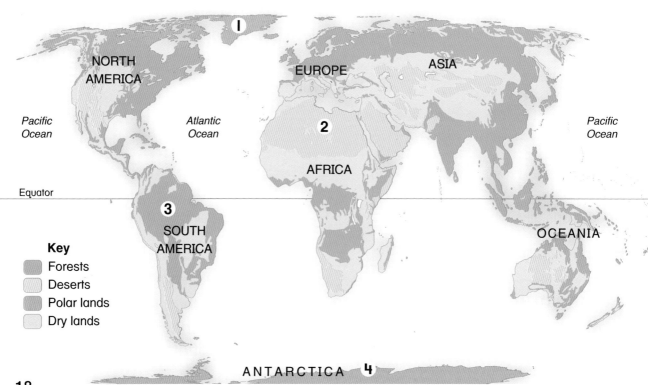

Unit 3 • Weather worldwide

Closer to home

In the UK we sometimes get hot, dry summer weather. The ground dries out and cracks appear. Rivers start to dry up. In the desert the weather is hot and dry nearly all the time.

In winter we sometimes get very cold weather. Ponds and lakes freeze over. Snow covers the land. Plants stop growing and animals hibernate. In polar lands the weather is like this even in summer.

▲ A dry reservoir in northern England.

▲ Snowy park in London.

Discussion
- How would you describe the weather where you live?
- Why do gardeners like to use greenhouses?
- Why are there no deserts or ice caps in the UK?

Mapwork
Match the numbers from the map to the places in this list.

Sahara Desert • Antarctica • Amazon rainforest • Greenland

Data Bank
- Highest ever UK temperature: SE England, 38° C.
- Lowest ever UK temperature: Scotland, -27° C.

Investigation
Where are the hardest climate conditions for plants and creatures in your school grounds?

Summary
In this unit you have learnt:
- that some parts of the world are always hot and other parts are always cold
- how to describe the weather in words and diagrams
- how the weather affects animals, plants and people.

Unit 4 Villages

Lesson 1: A village community

What makes a village?

A village is a place where a group of people have made their homes close together. If people start living on other planets they will have to build places where they can live safely. They will need shelter, food and water. In the past, villages were built here on Earth for the same reasons.

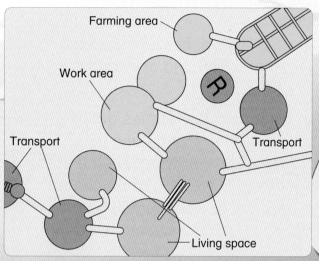

▲ Land use map for a settlement on Mars.

Key words

community
land use
planet
settlement
transport
village

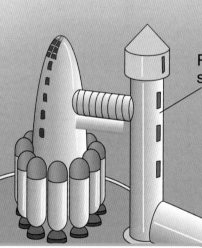

Discussion

☐ Which do you think are the three most important areas of the Mars village? Say why.

☐ Can you put the areas of the Mars village in order from the most to least important?

☐ If you lived on Mars which of the village jobs would you like to do?

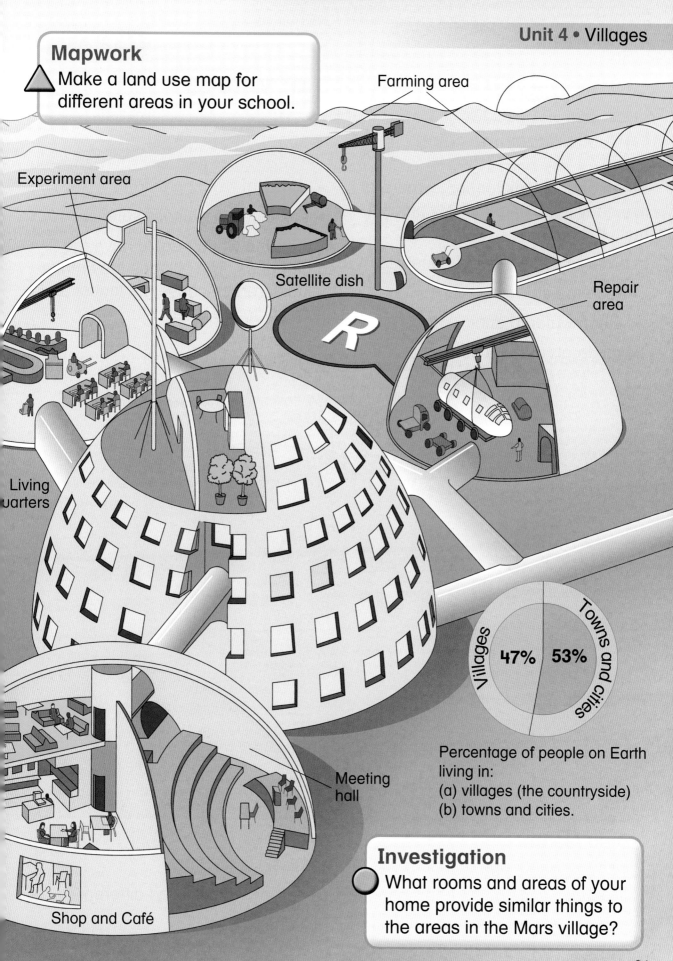

Unit 4 • Villages

Lesson 2: Different types of village

Key words
Alps
crops
desert
flood
materials

Are all villages the same?

When people look for somewhere to live, they try to find the best place to build their homes.

In order to survive people need:
- food to eat
- clean water to drink
- materials to build houses
- a way of keeping warm
- somewhere which is safe.

All over the world people have the same needs. The villages they make look different because the landscape and climate are different.

▼ Bainbridge, North Yorkshire.

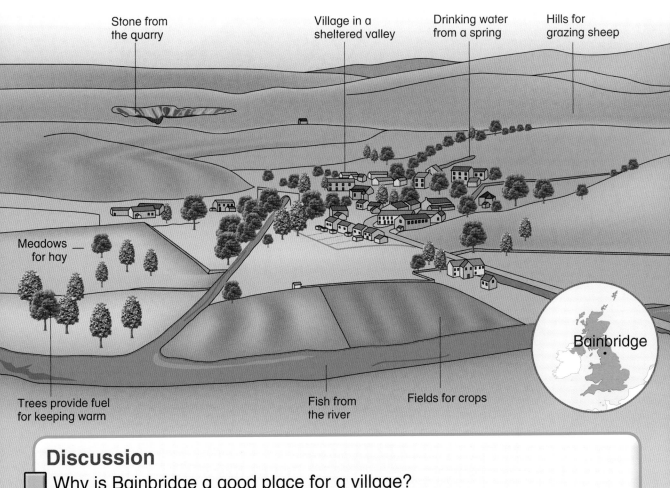

Stone from the quarry · Village in a sheltered valley · Drinking water from a spring · Hills for grazing sheep · Meadows for hay · Trees provide fuel for keeping warm · Fish from the river · Fields for crops · Bainbridge

Discussion
- Why is Bainbridge a good place for a village?
- In what ways are the houses in the photographs similar and different?
- Which village would you most like to visit and why?

Unit 4 • Villages

A desert village
Burkina Faso, West Africa

This village is on the edge of the Sahara Desert. The houses give shade and shelter from the sun. They have been grouped together for safety. The walls are made from baked earth and the roofs from dried grass. People keep goats and cattle.

A lake village
Myanmar, Southeast Asia

The people in this village make their living by fishing. They eat some of the fish and sell the rest in the market. The wood for their houses comes from the forest nearby. The houses are built on stilts to keep the village safe from floods.

A mountain village
Austria, Central Europe

This village is high in the Alps. The winters are cold and it often snows. The houses are made of stone and wood. Their roofs hang down over the walls to shed the snow. Some people in the village keep cows. The cows feed in the high meadows above the village in the summer.

Mapwork
Working from a local map, make a list of villages in your area or region.

Investigation
Draw pictures of the three villages and display them around a world map to show their location.

23

Unit 4 • Villages

Lesson 3: Investigating villages

How do villages change?

Key words
bungalow
church
greenhouse
orchard
paddock

1 Inn 2 Old farmhouse 3 School 4 Pond 5 Church

Apple orchard   Horse paddock   Field of Brussels sprouts

Worth is a village in southeast England. Two hundred years ago there was just a group of buildings around a church. In Victorian times new houses were added and a school was built.

Worth now has a small shop, a playing field and many more houses. A farmer has built some large greenhouses for tomatoes. The village is bigger than before but still has farmland all around it.

Many English villages are like Worth. They have grown larger as people who work in the towns move to the country.

Discussion
- What are the oldest parts of a village?
- How has Worth changed over the last two hundred years?
- What new feature would you add to Worth?

Unit 4 • Villages

Data Bank
- Most of the villages in England were founded before the Norman invasion of 1066.
- Villages vary in size from less than a hundred people to several thousand or more.
- Some villages have just one purpose such as fishing or mining.

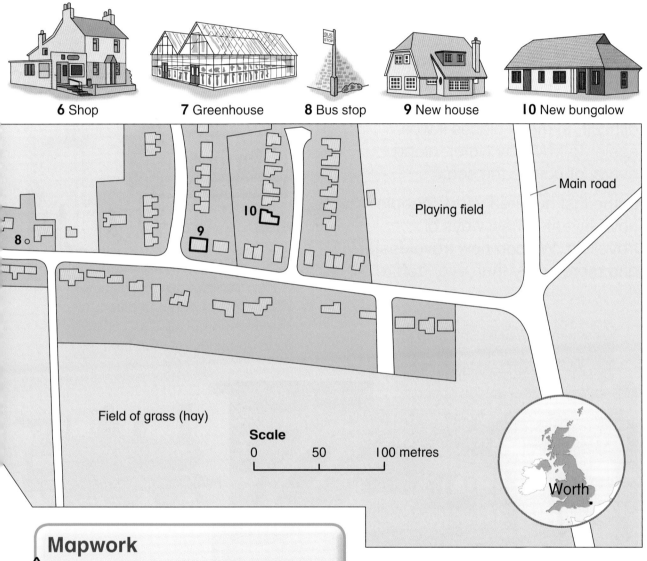

6 Shop   **7** Greenhouse   **8** Bus stop   **9** New house   **10** New bungalow

Mapwork
Draw pictures and plans of five buildings in Worth.

Investigation
Explore a village for yourself. Record what you find in photographs, maps and drawings.

Summary
In this unit you have learnt:
- what a village is
- about different villages around the world
- how to study a village.

Unit 5 Travel

Lesson 1: Ways of travelling

What different types of transport are there?

In the past, horses and carts helped people to move from place to place. Sometimes thick forests and high mountains made travel difficult. In many places it was easier to travel by water, along rivers, or across the sea.

In the last hundred years, people have invented new ways of travelling. We can now travel faster and more easily than ever before.

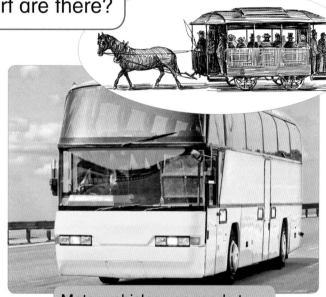

Motor vehicles use roads to reach lots of different places. They need garages for repair and petrol stations for fuel.

A single train can carry a lot of people or hundreds of tonnes of goods. Trains follow metal tracks and can travel at high speeds.

Discussion
- What are the main ways of travelling?
- How has transport changed?
- Which vehicle would you like to drive and why?

Key words
airport
harbour
transport
vehicle

Unit 5 • Travel

Ships and boats take people, vehicles and goods across water. They need harbours or sheltered places where they can tie up and unload.

Aeroplanes are the fastest way of travelling. They can cross mountains and seas without difficulty. They need airports for taking off and landing.

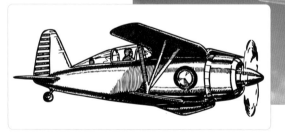

Data Bank
- Half of primary school children in the UK walk to school.
- There are 34 million vehicles in Britain.
- Around 180 million people use UK airports each year.

Mapwork
Make a timeline with drawings showing different ways of travelling over the last hundred years.

Investigation
Make a survey of how children in your class (a) travel to school (b) travel about at weekends.

Unit 5 • Travel

Lesson 2: Finding your way

Why do people use maps?

When people travel to a place for the first time they need to think which way to go. A map helps them to plan their route.

There are many types of map. Some show the centre of a town. They are drawn to a large scale and show a lot of detail. Others mark railway lines. They are drawn to a smaller scale and are not as detailed.

People who are driving, walking or travelling by train look for the map which shows their route best.

▲ Route from a house on the main road to the local church.

Discussion
- Why do people need different types of map?
- When do you use a map?
- Which of the maps on the opposite page do you think is most useful?

Van driver
When I am delivering goods to schools, I use a Satnav. This map shows street names, roundabouts and other landmarks.

Walker
When I am planning a walking trip, I use a map which marks the hills, footpaths, bridges and villages.

Train passenger
When I am travelling by train, I look at the station map to see which way to go.

Unit 5 • Travel

Key words

grid
landmark
route
scale

Street maps

A grid of small squares makes it easier to find places on a street map.

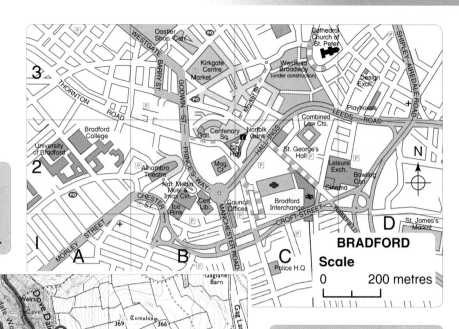

Footpath maps

Many footpath maps are made by Ordnance Survey. They show buildings and the shape of the land.

Train maps

Train maps show rail routes between places.

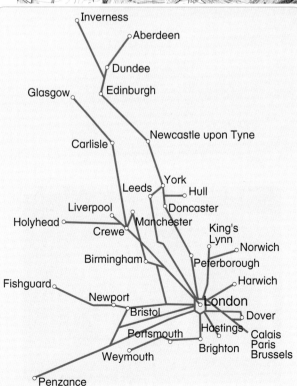

Mapwork

Draw a map to show a visitor how to reach your class from the school entrance.

Investigation

Make a class display of different maps of your area, the UK or wider word.

Unit 5 • Travel

Lesson 3: Routes and journeys

Key words
journey
landmark
Ordnance Survey
routes

Do routes matter?

Burydale Junior School is in Stevenage in Hertfordshire. Each week the children go by bus to the swimming pool in the middle of the town.

Normally the driver takes the route past Fairlands Valley Park. One week the driver had to go a different way because of road works.

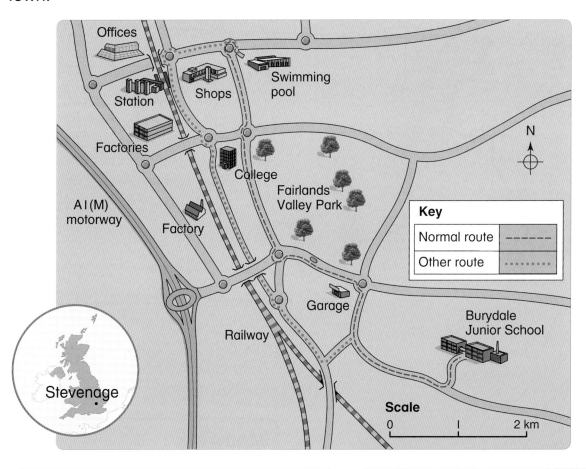

Discussion

- What landmarks did they pass on the new route to the swimming pool?
- What landmarks are shown on the child's map opposite?
- Why do you sometimes take different routes to the same place?

Unit 5 • Travel

As part of their project on routes the children talked about their journeys from home to school.

They made a list of the landmarks which they passed and then drew maps of their journeys.

Data Bank
- The oldest maps in the world are cave paintings made thousands of years ago.
- Modern maps use computers to help store and sort data.
- In Britain, Ordnance Survey maps were first made for the army to use in times of war.

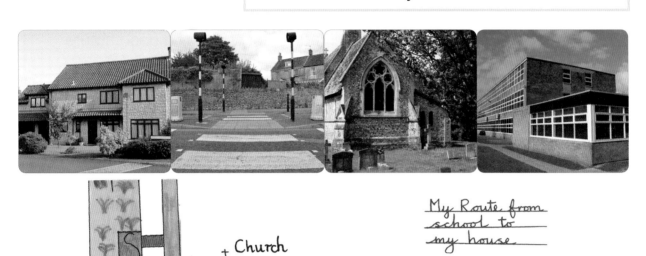

Mapwork
Make a list of the landmarks you pass on your way from home to school. Show them on your own route map.

Investigation
Find out how children in your class travel to school. Display the results on a bar chart.

Summary
In this unit you have learnt:
- what a route is
- how people use maps for journeys
- why landmarks are important.

Unit 6 Caring for the countryside

Lesson 1: Wildlife around us

What is a habitat?

The place where a community of plants and animals live is called a habitat. Ponds, woods, hedges, fields, waste ground and old walls are examples of different habitats. They provide food, water and shelter for the many plants and animals which live there.

Discussion

- Where might all the creatures that are looking for a home find a place to live in the old wall?
- What makes the old wall a good nature habitat?
- Can you think of any parts of your school grounds which are a bit wild?

Data Bank

- The number of otters, red kites and some rare types of butterfly is increasing in the UK.
- There are ten national parks in England, three in Wales and two in Scotland.

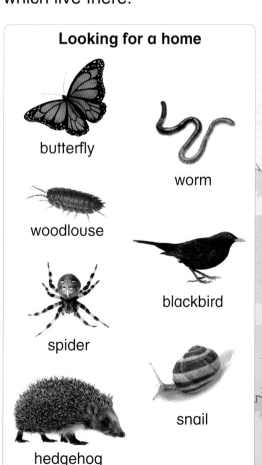

Looking for a home: butterfly, worm, woodlouse, blackbird, spider, snail, hedgehog

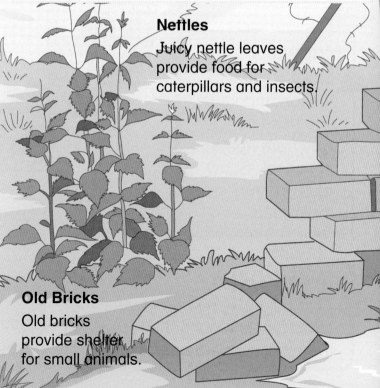

Nettles
Juicy nettle leaves provide food for caterpillars and insects.

Old Bricks
Old bricks provide shelter for small animals.

Unit 6 • Caring for the countryside

Key words

block graph
colour code
community
habitat
national park
survey

Branches
Trees have strong branches for nests.

Mapwork
Colour code a map of your school grounds. Use green for areas that are good for wildlife and yellow for areas where it is more difficult for wildlife to survive.

Buddleia
Butterflies feed on the nectar in flowers on buddleia bushes.

Tree trunk
Cracks in the bark make good homes for small animals.

Top of the wall
At the top of the wall there are dry cracks which can get very hot in the sunshine.

Bottom of the wall
The bottom of the wall is damp and shady.

Earth bank
Some animals live in the soil. Most plants need earth to grow.

Investigation
Make a survey of the front gardens in a street near your school. Decide if each garden is mostly grass, mostly plants and flowers or mostly covered in bricks, stone and concrete. Make a block graph of your results.

Puddle
Puddles provide water for animals to drink.

Unit 6 • Caring for the countryside

Lesson 2: **Protecting wildlife**

What are people doing to care for plants and animals?

All over the world people are trying to protect the environment. They want to look after the land and keep the air and sea clean.

In some places, people are protesting about pollution. New laws also help to protect the environment. However, it costs a lot of money to look after plants and animals and save their habitats.

Data Bank
- National parks cover 6% of the Earth's surface.
- Wildlife groups want 40% of the world's oceans to be made into reserves.

Yellowstone National Park, USA
Some beautiful landscapes have been turned into national parks for people to enjoy.

Colombia
Scientists have special areas where they can study wild plants. If nothing is done to save plants, they will be lost for ever.

Antarctica
Antarctica could very easily be spoilt by pollution. All countries have now agreed Antarctica should stay as a wilderness.

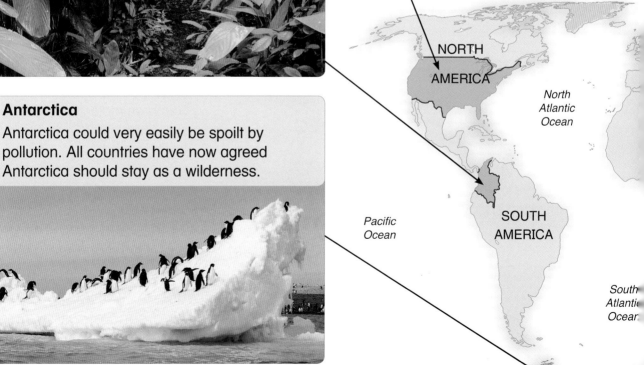

Unit 6 • Caring for the countryside

Discussion

- Why is it important to protect the environment around the world?
- Which conservation project do you think is most important?
- What could you do to protect wildlife?

Key words

conservation, national park, environment, pollution, habitats, reserve

Mapwork

Make a map of your nearest national park or conservation area.

Investigation

Devise a poster to make people want to care for wildlife.

Kenya
Game reserves have been set up to protect lions, elephants and other animals.

Northwest China
Thousands of trees have been planted to stop the desert from spreading over farmland.

Southern Ocean
People are trying to save whales from hunting which could make them extinct.

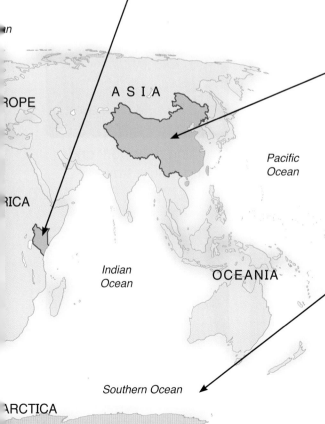

Unit 6 • Caring for the countryside

Lesson 3: Improving our surroundings

What happens at a nature reserve?

Key words

habitat
nature reserve
nature trail
warden

Andrew Jackson is the warden of a nature reserve. The reserve covers a large wood. Many birds, plants, butterflies and other insects live among the trees.

Andrew is always busy working in the reserve. These are some of the jobs that he does every year.

Discussion

- [ ] Why do we need nature reserves?
- [ ] What jobs does the warden do?
- [ ] What would happen if Andrew stopped looking after the reserve?

◀ Heath fritillary butterfly.

◀ Greater spotted woodpecker.

Autumn and winter

Cutting down old trees to make new homes for birds and insects.

Planting new trees.

Planning new work and writing nature trails.

Spring and summer

Counting birds and butterflies on the reserve.

Talking to visitors on walks.

Clearing spaces for flowers and butterflies.

Unit 6 • Caring for the countryside

Improving the school grounds

At Byron School the children joined a nature club. The club helped them to make their school grounds a better habitat for plants and animals.

Data Bank
- Around the UK, gardens are important habitats for wildlife.

▲ The children designed and made a vegetable garden.

▲ The children put food out to attract the birds.

▲ The garden is used for school project work.

Mapwork
Draw a plan of your school garden, or design a garden plan of your own.

Investigation
Devise a short nature trail for your school grounds.

Summary
In this unit you have learnt:
- what makes a habitat
- how people are caring for habitats in Britain and in other parts of the world.

Unit 7 Scotland

Lesson 1: Introducing Scotland

What is Scotland like?

Landscape

Many parts of Scotland have high mountains. There are lowlands in the middle of Scotland. The west coast has lots of islands.

Weather

In the Scottish mountains it is often wet and cold. During the winter, snow can cover the ground for many months at a time.

I live in Fort William. It rains on over 250 days in the year.

Aberdeen has less than half the rainfall of Fort William.

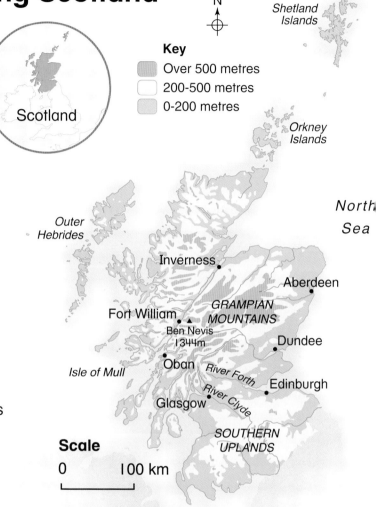

▲ Scotland is the most northerly part of the United Kingdom.

▲ Autumn in the Scottish mountains.

Unit 7 • Scotland

Key words

Ben Nevis
Edinburgh
Glasgow
River Clyde
Shetland Islands

Mapwork

List the islands shown on the map. Add others to your list using an atlas.

Discussion

- What three things would you tell a visitor about Scotland?
- How is Scotland different to where you live?
- Where do you think you would most like to live in Scotland?

Transport

The lowlands and east coast have the main road and rail routes. Ferries deliver food and goods to the islands.

Settlement

Edinburgh and Glasgow are the largest cities. Aberdeen and Dundee are two ports on the east coast. Very few people live in the mountains.

Work

Many people work in factories, tourism, and the oil industry. Farming, fishing and forestry are also important. Scotland is famous for tartan cloth and whisky.

▲ There are many rigs in the North Sea producing oil and gas.

▲ George Square, Glasgow.

Investigation

Using the internet find out more about (a) tourism (b) the oil industry in Scotland.

Unit 7 • Scotland

Lesson 2: Edinburgh: The capital city of Scotland

What is Edinburgh like?

Edinburgh is a very old city. A castle was built there hundreds of years ago. It stands on a high rocky crag above the city.

Today Edinburgh is a busy place with lots of traffic. People come a long way to go shopping in the city. Tourists arrive by plane, train and car. Lots of people work in the shops, offices and factories.

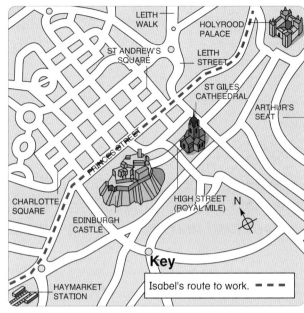

▲ A street map of central Edinburgh.

Discussion

- How can you tell Edinburgh has been an important city for a long time?
- How is Edinburgh different to the place where you live?
- Think of five places in Edinburgh where people might find work.

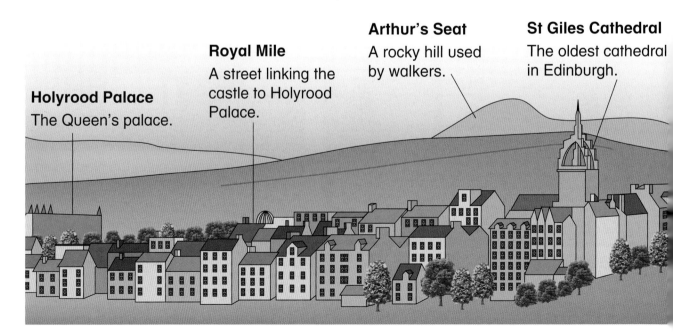

Holyrood Palace
The Queen's palace.

Royal Mile
A street linking the castle to Holyrood Palace.

Arthur's Seat
A rocky hill used by walkers.

St Giles Cathedral
The oldest cathedral in Edinburgh.

Unit 7 • Scotland

▲ A road and a rail bridge cross the Firth of Forth. The bridges are an important link between Edinburgh and other parts of Scotland.

Going to work

Isabel Andrews travels across Edinburgh every day on her way to work in a bank. The first thing Isabel passes is Haymarket station. Next she sees the castle. Isabel travels past the big shops on Princes Street. As she arrives at work she can see a rocky hill called Arthur's Seat. Millions of years ago this was a volcano.

Mapwork

Make a list of main landmarks which Isabel passes on her way to work.

Investigation

Using the internet find out about three places you would like to visit in Edinburgh.

Edinburgh Castle
The castle used to protect the city from attacks.

Key words

bank
castle
cathedral
crag
station

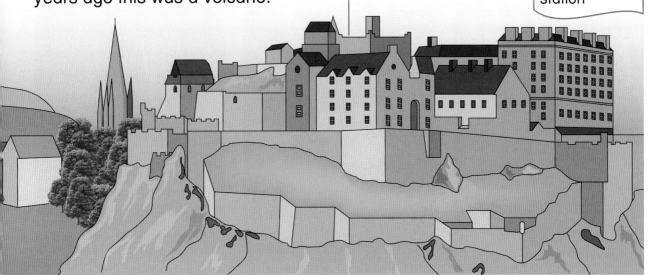

Unit 7 • Scotland

Lesson 3: Mull: A Scottish island

Key words

cliff
croft
ferry
island
moor

▲ Colourful houses line the harbour at Tobermory, the main town on Mull.

Living on Mull

Iain and Morag live on the Isle of Mull. Mull is a very quiet island in Western Scotland. There are mountains and moors in the centre and cliffs along the shore.

Iain and Morag run a small farm called a croft. Iain looks after the sheep and cows on the farm. Morag knits beautiful jumpers. She sells them to tourists who come to stay on the farm.

Key
main road —— ferry route - - -

▲ Mull is 48 kilometres from west to east and 42 kilometres from north to south.

Discussion

- What is the landscape like on Mull?
- What jobs are there to do on Mull?
- What do you think you would like and dislike about living on Mull?

▲ The old stone farmhouse.

▲ The farm animals.

42

Unit 7 • Scotland

Visiting Mull

Roy and Christine decided to go to Mull for their holidays because they enjoy bird watching. They arrived by ferry from Oban. The ferry was loaded with visitors, food and goods for the local people.

Data Bank
- About 4000 people live on the Isle of Mull.
- There are six primary schools and one secondary school.
- Over 200 types of bird have been recorded on Mull.

▲ View of Lismore lighthouse from Mull.

▲ Here are some of the birds Roy and Christine saw on their walks.

Christine bought a map so they could explore. Roy made sure they had waterproof anoraks in case it rained. They always took a picnic with them as there are very few shops outside Tobermory.

Mapwork
Plan a holiday which would take you to three Scottish islands. Draw a map of your route.

Investigation
Find out about the different things you can see and do in Western Scotland.

Summary
In this unit you have learnt:
- what makes Scotland special
- about the capital city of Scotland
- about a Scottish Island.

Unit 8 France

Lesson 1: Introducing France

What is France like?

France is about twice as large as the United Kingdom. It lies to the south of the United Kingdom across the English Channel.

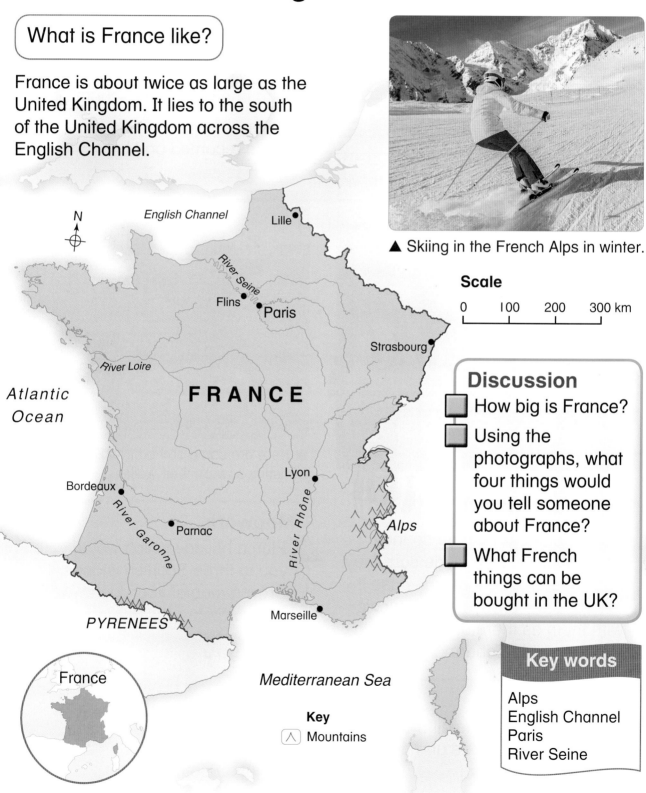

▲ Skiing in the French Alps in winter.

Discussion

- How big is France?
- Using the photographs, what four things would you tell someone about France?
- What French things can be bought in the UK?

Key words

Alps
English Channel
Paris
River Seine

44

Unit 8 • France

Landscape

The highest mountains in France are the Alps and Pyrenees. Their peaks are covered in snow and ice all year. The Seine, Loire, Garonne and Rhône are the longest rivers.

Weather

The north and west of France is often mild and quite wet. The south has hot, dry summers.

Work

France has many industries such as iron and steel, glass and chemicals, cars and aeroplanes and making clothes. People also work on the land. They produce fine wines, cheese, fruit, meat and vegetables.

▼ Many kinds of cheese are made in France.

▲ Nôtre Dame and the River Seine, Paris.

Settlement

Paris is the capital city. Other large cities are Lyon, Marseille, Bordeaux, Lille and Strasbourg. There are many country areas with scattered villages.

▲ High speed trains can reach speeds of 300 km an hour.

Transport

France has one of the most modern transport systems in the world. There are high speed railway lines and motorways.

Mapwork

Using an atlas draw a map of ferry routes between France and the UK.

Investigation

Make a list of all the different ways your class or community is linked to France.

Unit 8 • France

Lesson 2: Growing food

What crops do French farmers grow?

Parnac is a village in the Dordogne in the south-west of France. It has a church, old stone buildings and newer houses. The River Lot flows on one side of the village. Parnac has plenty of sunshine and the soil in the river valley is very good. Many crops are grown in the area.

Grapes are the most important crop. They are made into wine which is sold all over the world.

Some people keep geese and ducks. A type of meat paste, called pâté, is made from them.

Plums, apricots, apples and peaches grow in the orchards.

Tomatoes, beans and maize are grown for food.

Unit 8 • France

Key words

crops
export
settlement
soil
village

Discussion

- Why is Parnac a good place for a village settlement?
- What jobs do people do in Parnac?
- What do you think might make Parnac change?

Data Bank

- France produces more wine than any other country in the world.
- There are at least 350 different types of French cheese (one for every day of the year).
- France exports more food than any other country in Europe.

There are lots of caves under the ground. Thousands of years ago people used to paint pictures on the walls. Today tourists go to see the famous paintings.

There are many fine old country houses called châteaux in the Dordogne.

The milk from goats is used to make cheese.

The nuts from walnut trees are used in cooking.

Mapwork

Use an atlas to find out approximately how far it is from London to Bordeaux (the city closest to Parnac).

Investigation

Make a list of the food that is grown at Parnac. Discuss if each one is also grown in the UK. Show your answers with a tick.

Unit 8 • France

Lesson 3: Making cars

Key words
Europe, river, factory, settlement, motorway

> Where do Renault cars come from?

Renault is an important car company. It has factories in France and other countries across Europe. One of the oldest is at Flins about 40 kilometres from Paris.

Flins is near a motorway, railway and the River Seine. It is easy to reach by road and rail. There is also plenty of water. A lot of the people who work at the factory live in nearby towns and villages.

▲ Over 17 million vehicles, including the Renault Clio, have been made at Flins.

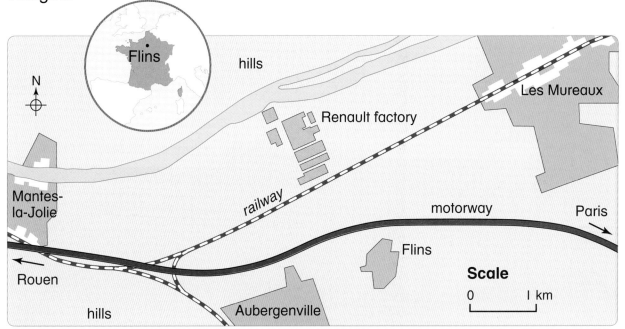

Data Bank
- The Flins car factory opened in 1952.
- It has produced 17 million cars.
- 3500 people work at the factory.

Discussion
- Why is Flins a good place for a car factory?
- What jobs do people do at Flins?
- What might happen to people in Flins if the factory closed?

Unit 8 • France

Working in Flins

Monsieur Hugo sprays the cars with paint. He travels from Mantes-la-Jolie each day.

Madame Renard lives in Les Mureaux. She checks that the cars are safe.

Monsieur Hassan puts the engines into the cars. He lives in Aubergenville.

Madame Blanc manages the supermarket near the factory. She lives in Flins.

▲ The city hall in Flins.

How the money goes round

People earn money in the factory.

They buy food at Madame Blanc's shop.

When Madame Blanc buys a Clio it helps to make more jobs.

▲ Flats in Aubergenville.

Investigation
Make a survey of parked cars in or around your school. How many of the cars are made by Renault?

Mapwork
Look at the map and list the settlements around Flins, the people who live there and the jobs they do.

Summary
In this unit you have learnt about:
- the landscape and weather of France
- french food and crops
- car making in France.

Unit 9 South America

Lesson 1: Introducing South America

What is South America like?

South America stretches southwards from the Panama Canal to Antarctica. The Andes mountains run down the western edge of the continent. There are many snowy peaks and active volcanoes. The River Amazon rises here. It flows east for 6440 kilometres before it enters the Atlantic Ocean at the Equator.

South America is divided into 13 countries. Brazil is much bigger than any of the others and covers half the continent. Five hundred years ago sailors from Spain and Portugal came to South America looking for gold and silver. Spanish and Portuguese are still the main languages in South America today.

Discussion

- What are the Andes?
- Why are European languages spoken in South America?
- What different things do the photographs tell you about South America?

▲ The modern part of Buenos Aires, Argentina, has wide, tree-lined avenues.

▲ Bolivian woman on a mountain trail by Lake Titicaca.

Unit 9 • South America

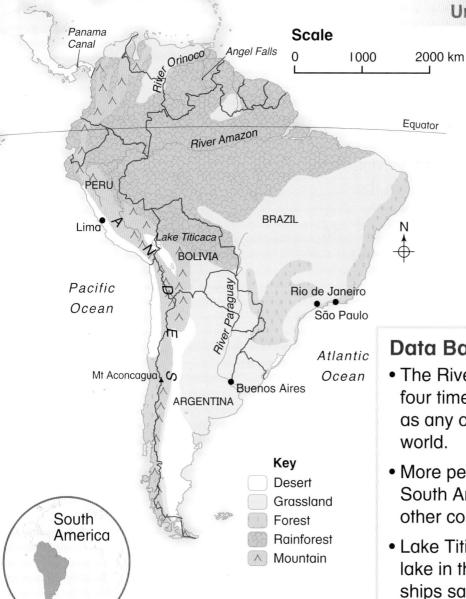

Key words

Andes
Brazil
Equator
Lake Titicaca
Panama Canal
Rio de Janeiro
River Amazon

Data Bank

- The River Amazon carries four times as much water as any other river in the world.
- More people live in cities in South America than in any other continent.
- Lake Titicaca is the highest lake in the world on which ships sail (3,811 metres).

Mapwork

Make a list of countries in South America which are (a) north of the Equator (b) south of the Equator (c) on the Equator.

Investigation

Draw your own map of South America. Mark six places or features. Write a sentence about each one.

▲ A tributary of the River Amazon winding through the rainforest.

Unit 9 • South America

Lesson 2: Spotlight on Chile

Key words
copper, glacier, desert, hot springs, fjord, salmon, geysers, volcano

What is Chile like?

Chile is shaped like a long, thin ribbon. It stretches over 4000 km down the coast of South America but is only around 200 km wide. This makes it one of the most unusual countries in the world.

Chile is a country of contrasts. The Atacama Desert in the north is one of the driest places on Earth. In the central region around the capital, Santiago, there is good farmland. This is the most crowded part of Chile. In the south, forests and glaciers reach down to the sea. Here there are flooded valleys called fjords similar to those in Norway.

Discussion
- What is special about the shape of Chile?
- What are the main regions in Chile?
- What three things do you think are most interesting about Chile?

Data Bank
- Chile is about three times the size of the UK. When it is winter in the UK it is summer in Chile.
- Santiago suffers from air pollution which becomes trapped by the Andes.
- There are geysers, hot springs and craters in the Atacama Desert.

Unit 9 • South America

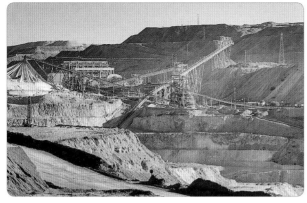

1 Copper mines

The Escondida mine in the Atacama Desert produces a third of the world's copper and employs 2500 workers.

3 Farming

Grapes grow well in central Chile. They are either made into wine or sent to supermarkets.

5 Salmon

The fjords in southern Chile are ideal for salmon. Chile is the world's biggest salmon producer after Norway.

Mapwork
Make simple outline drawings to show the shape of Chile, Norway and Vietnam.

2 Volcanoes

There are many volcanoes in the Andes. Lascar is one of the most active and has erupted many times.

4 Traditional crafts

Some people make traditional toys, gifts and clothes. Special deals ensure workers get a fair price for their goods.

Investigation
Make up a quiz. Write three sentences about Chile and three sentences about the UK. Mix them up and see if someone else can guess the country.

Unit 9 • South America

Lesson 3: The Galapagos Islands

Key words
Equator
ocean current
summit
volcano
heritage site

What is special about the Galapagos Islands?

The Galapagos are a group of remote islands about 1000 km from South America. Over thousands of years many different creatures have evolved in the Galapagos. Some of these cannot be found anywhere else in the world.

The Galapagos Islands were made into a World Heritage Site in 2001. Visitor numbers are strictly controlled. If new plants and creatures are brought from the mainland they could upset the balance of life.

▲ The Galapagos Islands.

Discussion
- What makes the Galapagos Islands remote?
- What might upset the balance of life on the islands?
- Looking at the photographs, which plants and creatures do you think are most remarkable?

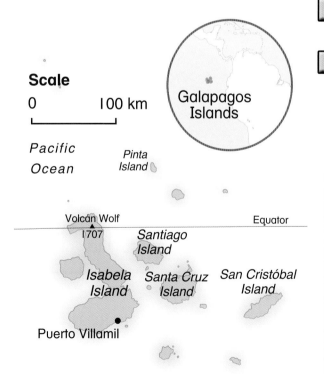

▲ There are over a hundred little islands in the Galapagos.

Data Bank
- Three ocean currents meet at the Galapagos Islands creating different ocean habitats.
- The Galapagos are the summit of an underwater volcano that rises 3000 metres from the ocean floor.
- 30,000 people live in the Galapagos Islands and there are 170,000 visitors each year.

Unit 9 • South America

Charles Darwin

Charles Darwin visited the Galapagos Islands in 1835 when he was a young man. He spent many years thinking about the remarkable plants and creatures he found there trying to explain why they had evolved. The theory that he came up with changed our ideas about life on Earth.

◀ Sally Lightfoot Crab.

▼ Frigate Bird.

▲ Giant Tortoise.

▲ Land Iguana.

Mapwork

Make a map of Isabela Island in your geography notebook. Write a sentence saying how far it is (a) across (b) round the coast.

Investigation

Make a small zigzag book about the Galapagos Islands and its wildlife.

Summary

In this unit you have learnt about:
- the features of South America
- what makes Chile special
- the plants and creatures of the Galapagos Islands.

Unit 10 Asia

Lesson 1: Introducing Asia

What is Asia like?

Asia is the largest continent. It is nearly 10,000 kilometres from east to west. In the north the climate is very cold. The south is hot with deserts and rainforests.

More than half the people in the world live in Asia. The largest countries are Russia, China and India.

Key words

China
Gobi Desert
Himalayas
India

Mapwork

Make a list of ten countries in Asia from an atlas.

▲ The Himalayas are the highest mountains in the world.

▲ Shanghai is one of the largest cities in China.

▲ The Gobi Desert is very dry and rocky.

Unit 10 • Asia

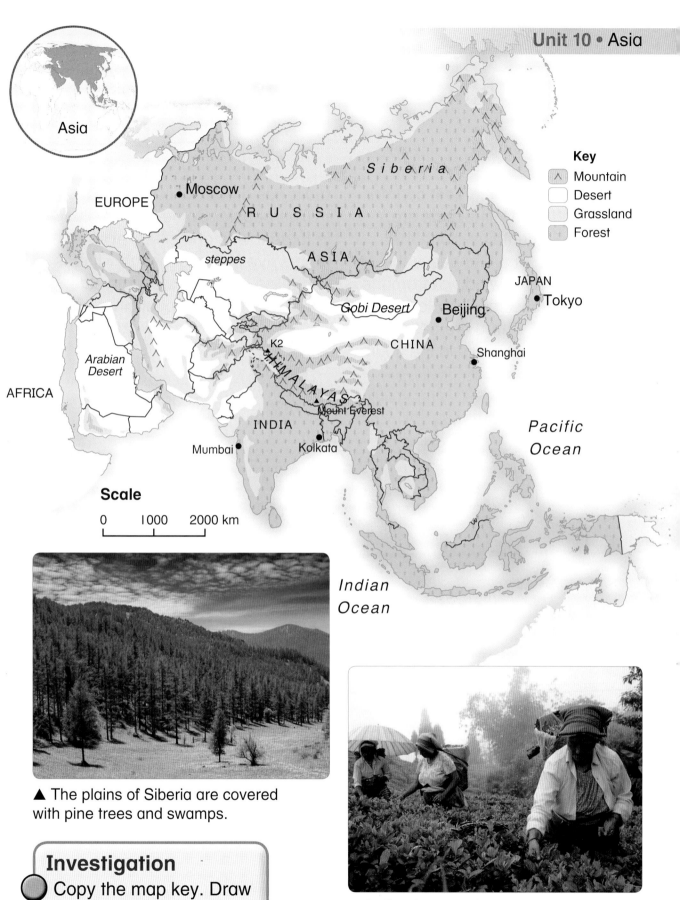

▲ The plains of Siberia are covered with pine trees and swamps.

▲ In Southeast Asia people depend on the monsoon rains to make their crops grow well.

Investigation

Copy the map key. Draw pictures to go with two of the landscape types.

Unit 10 • Asia

Lesson 2: India: A country in Asia

Key words
New Delhi
Himalayas
River Ganges
Thar Desert

What is India like?

India is the seventh largest country in the world. The Himalayas are a barrier of mountains along the north of India. The Ganges is the biggest river. It flows down from the mountains across a wide plain to the Bay of Bengal. There are deserts in the west and rocky hills in the south.

India has a population of 1200 million people. Some of them live in large towns and cities, and others live in thousands of villages scattered across the countryside. New Delhi is the capital city.

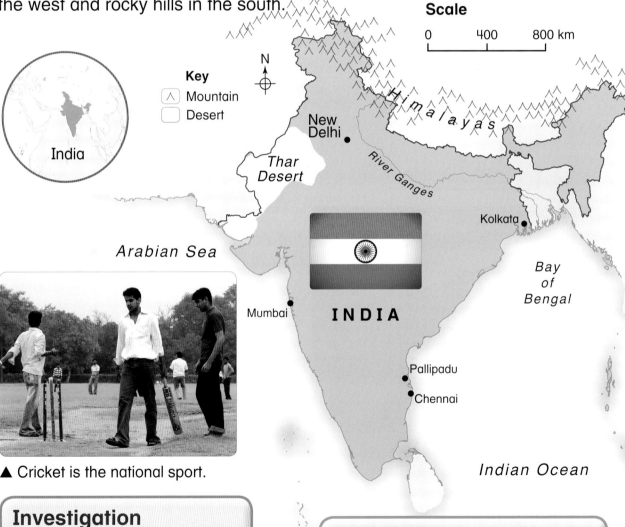

▲ Cricket is the national sport.

Investigation
Make a zigzag book about India. It could have a map, photographs and drawings.

Mapwork
Using an atlas make a list of countries which surround India.

58

Unit 10 • Asia

Lesson 3: Pallipadu: A village in India

What is life like in an Indian village?

"My name is Vijaya. I live in Pallipadu on the east coast of India. There are quiet streets and plenty of trees in my village. I think it is a beautiful place.

About 3000 people live in Pallipadu. Most of them are farmers. They grow rice, lentils, peanuts and vegetables.

▲ Vijaya

They also keep chickens and buffaloes. The soil is very fertile so people can grow two or three crops a year.

Between December and April the weather in Pallipadu is dry and warm. After that it becomes very hot and humid. When the monsoon rains come in July it is cooler.

Once there was a cyclone which caused terrible floods. Over a hundred people were drowned."

▲ Farmers use buffaloes to plough the fields.

Key words	
buffalo	monsoon
cyclone	rains
fertile soil	peanuts
flood bank	temple
lentils	wells

Jan	Feb	Mar	Apr	May	Jun	Jul	Aug	Sep	Oct	Nov	Dec
Pleasantly dry and warm			Hot and humid			Monsoon rains			Cyclones		

Unit 10 • Asia

"In the past all the water in Pallipadu used to come from wells and pumps. A few years ago the government built a water tank and put taps in the streets.

Buildings are changing too. The old houses had thatched roofs and mud walls. Now most of the houses are made of brick and cement. This means they are not so easily knocked down by cyclones. However, the new houses are small and they get hot inside in summer."

▲ A quiet side street.

Mapwork

Make a list and small drawings of the places shown on the map of Pallipadu.

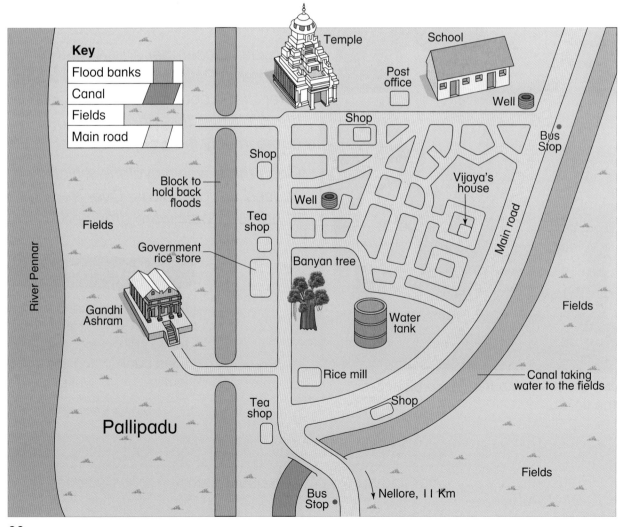

Unit 10 • Asia

"Lots of people in Pallipadu use motor scooters and bicycles. There is a bus which goes to the local town. People go there to sell vegetables or to see a film. At different times of the year there are festivals. We pray in the temple and afterwards there is dancing and everyone has a holiday."

▲ People enjoy special days like festivals and weddings.

▲ There are many small shops in all parts of India.

Discussion

- Which are the best months to visit Pallipadu?
- How is Pallipadu different to where you live?
- Do you think you would like to live in Pallipadu?

Investigation

Imagine you are visiting Pallipadu. Write an email to a friend saying what you saw on a tour of the village.

Summary

In this unit you have learnt about:

- the landscape of Asia
- different parts of India
- life in an Indian village.

Glossary

British Isles
Britain, Ireland and small islands around their coasts.

Climate
The pattern of weather over many years.

Continent
Great blocks of land, such as Africa.

Cyclone
A fierce storm which affects parts of Asia.

Desert
A dry area where there is very little rain.

Environment
The world around us.

Equator
An imaginary line around the Earth, half-way between the North and South Poles.

Cargo
The goods carried in vehicles like ships and planes.

Farm
A place where crops are grown and animals are kept for food.

Forestry
Growing trees so they can be sold for money.

Geography
The study of the surface of the earth and how people live.

Goods
The things which people sell to each other.

Habitat
The place where plants and animals live.

Hibernate
An animal hibernates when it goes to sleep during the winter months.

Landscape
The shape of the land, which can be made up of mountains, hills, valleys and other features.

Monsoon rains
Rainy weather which comes after the dry season in Southeast Asia.

North Pole
The most northerly point on the earth's surface.

Oasis
A place in the desert where there is enough water for trees and plants to grow.

Ocean
The seas which surround the continents.

Planet
A mass of rock and gas which circles around a star.

Plateau
A fairly flat piece of land high up in the mountains.

Polar lands
The land and ice around the north and south poles.

Pollution
Changes in our surroundings which damage the health of people, plants or animals.

Rainforest
Areas of thick forest which are near to the Equator.

Settlement
The places where people live, such as villages, towns and cities.

South Pole
The most southerly point on the earth's surface.

Tourist
Someone who visits places on holiday.

Transport
The vehicles used by people or goods.

United Kingdom
The country made up of England, Wales, Scotland and Northern Ireland.

Water vapour
An invisible gas in the air.

Index

aeroplanes 27
Africa 8, 23, 35
air 2, 8, 13
airports 27
Alps, mountains 23, 44, 45
America, North 34
America, South 4, 34, 50-55
Andes, mountains 4, 51, 52

animals 2, 10, 14, 15, 16, 19, 32, 33, 34, 35, 37, 42
Antarctica 14, 16, 34, 35, 40
Arctic Ocean 34, 35
Asia 4, 23, 35, 56-61
Atlantic Ocean 18, 44, 51
Austria 23

Bainbridge, Yorkshire 22
Ben Nevis 39
birds 10, 37, 43
Britain 27, 31, 37
British Isles 6, 42
Burkina Faso 23
butterflies 33, 36

China 35, 56
climate 14, 22, 51, 56
clouds 9, 13
coast 5
Colombia 34
Continents 2, 56
cows 23, 42
crops 16, 17, 57, 59

deserts 2, 14, 15, 16, 17, 18, 19, 35, 56, 58

Edinburgh, Scotland 39, 40-41
England 5, 19, 24, 25, 32
Equator 18, 51 55
Europe 18, 23, 35, 47, 48, 53

factories 48
farmland 5, 24, 35, 52
fishing 16, 23, 39
food 10, 16, 17, 20, 22, 25, 32, 37, 39, 43, 46, 47, 49
forests 2, 14, 15, 19, 23, 51, 52
France 5, 44-49

game reserves 35
Ganges, river 58
glaciers 9, 52
Gobi, desert 56, 57
goods 26, 27, 28, 39, 43, 53
Greenland 16, 19

habitats 32, 33, 34, 35, 36, 37, 54
hills 5, 6, 13, 28, 40, 48, 58

Himalayas 4, 56, 57, 58

ice 4, 9, 14, 15, 19, 45
icebergs 8, 9
India 56, 57, 58-61
Indian Ocean 35, 57, 58
industries 45, 39
insects 10, 32, 36
Ireland 6
islands 4, 5, 6, 12, 39, 39, 42, 43, 54, 55

Kenya 35

lakes 8, 9, 12, 13, 19, 23, 50, 51
landscape 2-7, 22, 34, 38, 42, 45, 49, 57, 61
Lot, river 48
lorries 16
lowlands 5, 6, 38, 39

maps 6, 12, 18, 24, 25, 28, 29, 30, 31, 34, 35, 38, 40, 42, 44, 48, 51, 52, 54, 57, 58, 60
marshes 5, 12
mountains 2, 4, 5, 6, 23, 26, 27, 38, 39, 41, 42, 45, 50, 56, 58

Mull, Scotland 42-43
Myanmar 23

national parks 32, 33, 34, 35
nature reserves 36
North Pole 14, 18

oasis 17
Oceania 35
oceans 2, 8, 34, 54
oil 39
orchards 24, 46

Pacific Ocean 2, 8, 18, 34, 35, 51, 52, 54
Pallipadu, India 59-61
Parnac, Dordogne 46-47
people 5, 10, 16, 17, 20, 21, 22, 23, 24, 26, 27, 28, 34, 35, 39, 40, 43, 45, 46, 47, 48, 49, 53, 56, 57, 58, 59, 61
plants 2, 4, 10, 11, 13, 14, 15, 19, 32, 33, 34, 36, 37, 54, 55
plateau 4
polar lands 14, 15, 16, 17, 18, 19
pollution 34, 35, 52
ponds 8, 10, 11, 13, 19, 24, 32
puddles 13, 33
Pyrenees, mountains 44, 45

railways 28, 30, 45, 48
rain 4, 9, 13, 14, 17, 38, 57, 59

rainforest 14, 15, 19, 51, 56
Renault 48-49
reservoirs 12
rice 59, 60
rivers 5, 6, 8, 9, 12, 13, 14, 19, 26, 39, 44, 45, 46, 48, 50, 51, 58
roads 25, 26, 39, 41, 48, 60
route 28, 29, 30, 39
rocks 2, 3, 6, 8
Russia 56, 57

Sahara 14, 19, 23
Scotland 19, 32, 38-43
sea 2, 5, 6, 8, 9, 12, 26, 27, 34, 52
Seine 44, 45, 48
settlements 20, 39, 45, 47, 48, 49
sheep 22, 42
ships 27, 51
shops 21, 24, 25, 30, 40, 41, 43, 49, 61
Siberia, Russia 57
snow 4, 9, 15, 19, 23, 38, 45
soil 4, 5, 8, 10, 13, 33, 46, 47, 59
South America
Southern Ocean 35
South Pole 18
streams 8, 12, 13
sun 2, 3, 17, 18, 23
swamps 5, 57

tourism 39
trains 26, 28, 29, 40, 45
transport 20, 26, 39, 45
travel 26-31
trees 4, 5, 10, 13, 14, 33, 36, 36, 47, 57, 59

United Kingdom 10, 11, 17, 19, 27, 29, 32, 44, 45, 47, 52, 53
United States 16, 34

valleys 5, 22, 46, 52
Victoria Falls, Africa 8
villages 17, 20-25, 28, 45, 46, 47, 48, 58, 59, 61

Wales 5, 32
water 2, 5, 8-13, 15, 17, 19, 20, 22, 26, 27, 32, 48, 53, 60
water vapour 9
weather 4, 14-19, 38, 45, 59
whales 35
wildlife 32-37
work 24, 39, 40, 41, 45, 48, 53
Worth, Kent 24-25

Zambezi 8

63

Primary Geography Book 3
Collins
An imprint of HarperCollins Publishers
Westerhill Road
Bishopbriggs
Glasgow
G64 2QT

© HarperCollins Publishers 2014
Maps © Collins Bartholomew 2014

© Stephen Scoffham, Colin Bridge 2014

The authors assert their moral right to be identified as the authors of this work.

ISBN 978-0-00-756359-3

Imp 001

Collins ® is a registered trademark of HarperCollins Publishers Ltd

All rights reserved. No part of this publication may be reproduced, stored in a retrieval system, or transmitted in any form or by any means, electronic, mechanical, photocopying, recording or otherwise, without the prior written permission of the publisher or copyright owners.

The contents of this edition of Primary Geography Book 3 are believed correct at the time of printing. Nevertheless the publishers can accept no responsibility for errors or omissions, changes in the detail given, or for any expense or loss thereby caused.

British Library Cataloguing in Publication Data
A catalogue record for this book is available from the British Library.

Printed and bound by South China Printing Company Ltd

Acknowledgements

Additional original input by Terry Jewson

Cover designs Steve Evans illustration and design

Illustrations by Jouve Pvt Ltd pp 4, 5, 10, 13, 17, 18, 20, 21, 22, 24, 25, 30, 32, 33, 40, 41, 60

This product uses map data licensed from Ordnance Survey® with the permission of the Controller of Her Majesty's Stationery Office.
© Crown copyright. Licence number 100018598

Photo credits:

(t = top b = bottom l = left r = right c = centre)

© Aerial Photography Solutions Ltd p28tr; © hagit berovich/Shutterstock.com p11tl;
© meunierd/Shutterstock.com p50bl; © nevenm/Shutterstock.com p11tr;
© Nicku/Shutterstock.com p55tl; © ostill/Shutterstock.com p45br;
© PaveSvoboda/Shutterstock.com p57br; © saika3p/Shutterstock.com p61l;
© Stephen Scoffham: p60; © TachePhoto/Shutterstock.com p48;
© Vibrant Image Studio/Shutterstock.com p26l; © zeber/Shutterstock.com p58bl

All other images from www.shutterstock.com